Ways Things Move

By Jennifer Kelley

Printed in Mexico

ISBN 978-0-15-365319-3
ISBN 0-15-365319-1

2 3 4 5 6 7 8 9 10 805 16 15 14 13 12 11 10 09 08

Harcourt
SCHOOL PUBLISHERS

Visit *The Learning Site!*
www.harcourtschool.com

ways things move

straight

in circles

back and forth

zigzag

curved

We move in different ways!